Baofeng Radio Survival Han

CONTENTS

Introduction

Overview of the Book

Welcome, reader, to a journey into the fascinating world of Baofeng radios and their critical role in survivalism, emergency preparedness, and in times of conflict. This book aims to provide an in-depth understanding of these robust devices, arming you with knowledge and practical skills to wield the power of radio communication effectively. As we delve into the technical aspects, the guide is designed to remain accessible to beginners, building your comprehension from foundational concepts to more complex operational procedures.

This book serves not just as a technical manual for Baofeng radios but as a comprehensive guide to survivalist communication. It is designed to help you understand the principles of radio communication, how to operate and program your Baofeng radio, how to use it in survival and emergency situations, and how to maintain and troubleshoot your device. Through these pages, we'll unravel the intricate web of knowledge surrounding Baofeng radios, explaining it in simple, easily understood terms.

Our journey begins with an exploration of basic radio communication principles, with the focus on why ham radio is an indispensable tool for survivalism and emergency preparedness. We will unravel how these devices function and what distinguishes them from other communication devices. This exploration will not only clarify the science behind radio communication but will also underline the value of understanding this medium in our increasingly digital, network-dependent world.

Once we've established this foundational understanding, we'll dive into the specifics of Baofeng radios. These devices come in various models, each with its unique set of features and capabilities. We will explore these models, highlight their critical features, and guide you through their basic operation and more advanced functionalities.

Having a Baofeng radio in your hands is one thing; knowing how to program it is quite another. We'll guide you through this process, offering detailed, step-by-step instructions for manual programming and programming through CHIRP software. We will explore the various settings you can customize to make your radio fit your specific needs and preferences.

The book then pivots towards survivalist communication, detailing the various protocols and practices crucial for utilizing your Baofeng radio in survival situations. Here, we delve into how to connect with emergency services and how to build a network with fellow survivalists. As with any piece of technology, proper maintenance and troubleshooting are vital. We will share best practices for caring for your radio and a guide to common problems and their solutions.

Finally, the book closes with real-life use cases and practical exercises to help you apply the knowledge you have accumulated throughout your reading journey. You will be exposed to scenarios that present the practical application of Baofeng radios in survival situations. This hands-on approach will cement your understanding and boost your confidence in operating these devices in real-world contexts.

At the end of this journey, our hope is that you'll have not only an understanding of Baofeng radios but also a profound appreciation for the world of radio communication. As we navigate these intricate waters, remember that every journey begins with a single step. It might feel daunting at first, but as you progress, your understanding will deepen, and the pieces will fall into place.

We invite you now to step into this captivating world of Baofeng radios. In the following chapter, we will discuss why Baofeng radios are such valuable tools in survival and emergency situations, providing a strong foundation for the rest of our journey. We look forward to navigating this journey together and uncovering the numerous ways in which Baofeng radios can be your lifeline when it matters most.

Why Baofeng for Survival and Emergency Situations

As we step further into the world of radio communication, it's crucial to recognize why Baofeng radios hold such a special place in survivalism and emergency preparedness. Throughout this chapter, we'll unravel the inherent qualities that make these radios a reliable and invaluable tool when faced with critical situations, focusing on their reliability, range, versatility, and affordability.

When disaster strikes or an emergency situation unfolds, conventional communication networks such as cell phones and the internet can falter or fail entirely. These services, as indispensable as they are in daily life, are ultimately dependent on infrastructural stability - something that is often the first casualty in any crisis. Here, the **reliability** of Baofeng radios shines through. These devices are designed to operate independently of conventional communication networks. As long as they are powered, they can transmit and receive signals, enabling you to maintain critical communication even when all else fails.

Next, we turn to the question of range. The ability to communicate across significant distances is paramount in many survival and emergency scenarios. The range of your Baofeng radio is influenced by several factors, including the specific model of your radio, the type of antenna used, and the frequency band on which you are operating. Under optimal conditions and with the right setup, a Baofeng radio can potentially reach other operators tens or even hundreds of miles away. This far-reaching capability makes it an invaluable tool for communication in remote or widespread areas, or in scenarios where moving around to find help isn't an option.

The **versatility** of Baofeng radios further solidifies their value in emergency situations. These devices can operate across a wide range of frequencies, including the Very High Frequency (VHF) and Ultra High Frequency (UHF) bands. This wide range of accessible frequencies allows you to tap into various channels, including those used by emergency services, local law enforcement, or other amateur radio operators. Such access can become a vital lifeline in an emergency, providing you with critical updates, facilitating calls for assistance, or simply maintaining a link to the outside world when isolation is a real and immediate threat.

Finally, the **affordability** of Baofeng radios cannot be overlooked. Despite their robust capabilities and reliable performance, Baofeng radios are surprisingly economical compared to many other brands in the market. This accessibility makes them a popular choice for many individuals interested in emergency preparedness, effectively democratizing the power of radio communication to a wide range of users, from professional first responders to everyday citizens.

Having a Baofeng radio and knowing how to use it effectively can dramatically increase your preparedness for a wide variety of emergency scenarios. This simple device can provide a lifeline in moments of crisis, facilitate coordination in survival situations, and offer reassurance through the simple act of maintaining a connection to the wider world.

As we continue our exploration in the next chapter, we will delve into the principles of radio communication, to further deepen your understanding of how these devices work and how to use them effectively. We'll learn about frequencies, bands, modulation, and more, setting the stage for our later discussions on operating and programming your Baofeng radio. So, hold tight as we dive deeper into the intricate and compelling world of Baofeng radios and their crucial role in survival and emergency situations.

Chapter 1: Understanding Radio Communication

Basic Radio Principles

As we embark on the next phase of our journey into the world of Baofeng radios, it's essential that we start by gaining a fundamental understanding of the underlying principles of radio communication. This knowledge will serve as the bedrock for our future exploration into the operation and programming of these devices. So let's dive into the basics of radio communication, starting with concepts like frequencies, bands, and radio waves.

At its heart, radio communication is the transmission and reception of information through **radio waves**. These are a type of electromagnetic wave, much like light waves, but they exist at a different point in the electromagnetic spectrum and exhibit unique properties that make them suitable for long-distance communication.

The **frequency** of a radio wave is a critical characteristic that determines many of its properties. Measured in Hertz (Hz), frequency refers to the number of cycles a wave completes in one second. The human ear can hear sounds within a certain frequency range, while radio devices can tune into radio waves within specific frequency ranges. Understanding frequency is crucial because each frequency carries its own set of regulations and restrictions, and knowing which frequencies to use (and which to avoid) can mean the difference between effective communication and radio silence.

Another vital term in radio communication is the **band**. Bands are simply ranges of frequencies. For example, the **Very High Frequency (VHF)** band includes frequencies from **30 to 300 MHz**, while the **Ultra High Frequency (UHF)** band ranges from **300 MHz to 3 GHz**. Baofeng radios typically can operate within the VHF and UHF bands, making them incredibly versatile tools for communication. It's important to understand the different bands and what they are used for because each band has its strengths and weaknesses, depending on the specific use case.

Radio communication also involves the principle of **modulation**, which is the process of transforming the voice or data that you want to transmit into a form that can be sent over the airwaves. There are several types of modulation, but the two most common types you'll encounter when using a Baofeng radio are Frequency Modulation (FM) and Amplitude Modulation (AM). FM is the most commonly used modulation for VHF and UHF frequencies due to its clarity and resistance to interference.

Lastly, understanding the concept of **antennas** is crucial in grasping radio communication principles. Antennas serve as the bridge between the electrical signals within your radio and the radio waves in the atmosphere. The size, shape, and type of the antenna can greatly affect your radio's performance.

These foundational concepts — radio waves, frequency, bands, modulation, and antennas — form the backbone of radio communication. With this fundamental understanding, you are well-equipped to delve deeper into the operational specifics of your Baofeng radio, the subject of our next chapter.

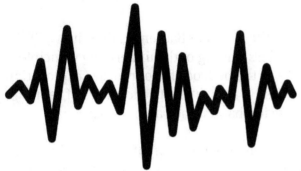

We will explore the various models and their unique features, providing you with the knowledge you need to choose the right Baofeng radio for your needs. So, as we close this chapter on basic radio principles, we open a new one on the fascinating intricacies of Baofeng radios themselves.

Why Ham Radio?

Having laid the foundation with the basic principles of radio communication, we can now delve into the fascinating world of amateur radio, commonly known as "ham radio." This subchapter will explain what ham radio is and explore its relevance to survivalism and emergency preparedness, showcasing why it is so highly valued in these contexts.

Ham radio, at its core, is a diverse and richly engaging hobby that also serves a vital role in emergency communications. Amateur radio operators, or "hams," use various types of radio communications equipment to communicate with other hams for public service, recreation, and self-training. But what distinguishes ham radio from other forms of radio communication?

One defining aspect is the **accessibility and diversity of communication modes** that ham radio offers. In addition to voice communication, hams can use Morse code, digital modes, and even image transmission methods to communicate. This array of options makes ham radio adaptable to a wide range of situations, providing a lifeline when traditional communication methods are unavailable.

The **self-sufficient nature** of ham radio is another critical factor that contributes to its relevance in survivalism and emergency preparedness. Ham radios operate independently of any external infrastructure. They don't rely on cell towers, internet connectivity, or the electrical grid to function, meaning they can provide reliable communication in a variety of adverse conditions when other methods fail.

Another key benefit is the **global reach** of ham radio. With the right equipment and conditions, you can communicate across town, around the world, or even into space. This range of communication expands potential sources of assistance and information in emergency situations, transcending geographical and political boundaries.

Moreover, the ham radio community places a high value on **public service and emergency readiness**. Hams often participate in emergency drills and provide crucial communication services during real-world disasters. As a ham radio operator, you'll be joining a community committed to lending a hand when it's most needed.

Finally, **education and experimentation** are fundamental aspects of ham radio. The very process of preparing for the license exam provides a basic education in radio technology and operating principles. It's a continual learning experience that equips you with knowledge and skills which can prove invaluable in emergency and survival scenarios.

Before we close this subchapter, it's important to note that operating a ham radio does require a license in most countries. The licensing process is there to ensure operators understand radio etiquette, technical fundamentals, and relevant legalities. However, in a dire survival situation where all other communication methods have failed, unlicensed operation in the spirit of immediate safety and preservation of life is generally accepted.

Understanding the power of ham radio, you're now poised to navigate the specifics of Baofeng radios in the upcoming chapters. We'll be discussing the different models, their functionalities, and how to operate them effectively. Having grasped why ham radio is such a crucial tool for survivalism and emergency preparedness, you'll see how Baofeng radios, as an accessible gateway into this world, can be your lifeline when you need it most. So, let's continue the journey together, stepping further into the fascinating realm of Baofeng radios and their role in survival and emergency situations.

Amateur Radio Licensing

As we move forward into the realm of amateur radio operation, it's vital to understand that one does not simply pick up a ham radio and start transmitting. There are rules and regulations to follow, a primary one being the requirement for an amateur radio license. This subchapter aims to guide you through the fundamentals of the licensing process for amateur radio operation and the steps to obtaining your own license.

In most countries, the operation of amateur radios is overseen by a regulatory body. In the United States, it's the Federal Communications Commission (FCC), while in the United Kingdom, it's Ofcom. These organizations are responsible for issuing licenses and ensuring compliance with the rules of radio communication.

An **amateur radio license** is a legal permission that grants its holder the privilege to operate and transmit on certain radio frequencies allocated for amateur use. It's a testament to the operator's understanding of radio operation principles, etiquette, and legal regulations. It's not meant to be a deterrent but a means to maintain order on the airwaves and prevent harmful interference.

Obtaining an amateur radio license usually involves passing an examination. The complexity of the exam typically corresponds with the level of the license. In the United States, there are three license levels: Technician, General, and Extra, each granting additional privileges on the spectrum. The **Technician license**, being the entry-level license, grants privileges on bands above 30 MHz and some limited privileges on the shortwave bands. It is the perfect starting point for newcomers to ham radio.

To prepare for the exam, there are numerous resources available, from study guides to online practice exams. These materials cover a variety of topics including basic radio theory, FCC regulations, operating practices, and safety. A common approach is to study a dedicated manual that's tailored for the particular license level you're aiming for, and then supplement your study with online practice tests. The aim is not merely to pass the exam but to gain the knowledge and skills required for responsible and effective operation of your radio equipment.

Once you feel confident in your understanding, you can schedule your examination. Exams are usually administered by Volunteer Examiners (VEs), who are licensed hams authorized by the regulatory body to conduct the exams. Depending on your location, exams can be taken in person at a designated testing center, or online.

Once you pass the examination, you'll receive your call sign from the regulatory body. This call sign is your unique identifier on the airwaves and should be used at the start and end of all transmissions, as well as periodically during longer transmissions.

While the process may seem daunting at first, earning your amateur radio license is a significant achievement that opens up a world of opportunities in the ham radio community. It's also a stepping stone to becoming a competent operator, a role that can prove vital in survival and emergency situations.

With the licensing process demystified, our journey will now venture into the specifics of Baofeng radios. From understanding different models to setting them up and programming them, we'll cover it all. You're well on your way to becoming a fully-fledged amateur radio operator, capable of leveraging the power of your Baofeng radio in any situation that may arise.

Chapter 2: Baofeng Radios

Baofeng Models and Features

Now that we have traversed the landscape of amateur radio and licensing, let's delve into the heart of this guide – Baofeng radios. Baofeng offers a variety of models, each with their unique features and specifications designed to meet various user needs. This subchapter provides a comparison of popular Baofeng models and their key features, helping you make an informed choice based on your requirements and preferences.

Baofeng UV-5R

Arguably the most well-known Baofeng radio, the UV-5R is a dual-band handheld transceiver operating on VHF (136-174 MHz) and UHF (400-520 MHz) frequencies. This radio is a favorite among beginners due to its affordability and simplicity. Despite its low price, it does not compromise on features, offering high/low power settings, a 128-channel memory, a built-in flashlight, and even an FM radio receiver. It also supports dual watch, which allows monitoring of two channels.

Baofeng BF-F8HP

The BF-F8HP is a step-up from the UV-5R, offering higher power output for better range. It has three power levels (low, medium, high) with the highest power setting delivering up to 8 watts. It maintains the dual-band functionality and shares the UV-5R's feature set while boasting an improved antenna, a better battery life, and a more detailed manual, making it a popular choice among more experienced users.

Baofeng UV-82HP

The UV-82HP is another high-powered Baofeng radio with an output of up to 8 watts. It stands out with its unique design that includes dual PTT (Push-to-Talk) buttons, enabling easy operation of its dual-watch capability. This model also features a more robust and comfortable build quality compared to the UV-5R, and it offers advanced options like group tones and selective calls.

Baofeng GT-3TP

The GT-3TP is a top-tier offering from Baofeng with power output up to 8 watts. Its standout features include a rugged, ergonomic design for enhanced durability and handling, a new chipset with noise-reduction features for clearer communication, and a high-gain antenna for improved range. The GT-3TP supports a full range of VHF and UHF frequencies and includes a robust feature set that caters to advanced users.

Baofeng UV-3R

For those who desire a more compact option, the UV-3R is a small, lightweight model that is simple and convenient to carry around. It operates with lower power (up to 2 watts) and has fewer channels (99), but it covers the essential features for basic communication needs. This radio is a great choice for individuals who prioritize portability and simplicity.

While this comparison highlights key differences among popular models, it's crucial to note that the best radio for you largely depends on your specific needs and circumstances. Consider factors such as your budget, power needs, desired features, and the radio's intended usage scenario. For instance, a prepper might prioritize a robust model with a high power output for potential emergency situations, while a hobbyist might prefer a model with advanced features for exploring different aspects of ham radio.
With this understanding of Baofeng models and their features, we can now explore the technicalities of setting up and programming your Baofeng radio.

The upcoming chapters will guide you through these steps, ensuring you're ready to operate your radio effectively when the need arises. As we continue, remember, each journey with a Baofeng radio is unique, and there's always more to learn and explore.

Basic Operation

As you embark on your journey with Baofeng radios, understanding the basic operation of these devices is the first step towards effective communication. In this subchapter, we'll demystify the fundamental operations, from powering the radio on and off, adjusting the volume, changing frequencies, to various other elementary operations that will equip you to use your Baofeng radio effectively.

Powering On and Off

Starting with the most basic operation, the power button on a Baofeng radio is usually combined with the volume knob at the top of the device. To turn the radio on, rotate the knob clockwise. You'll be greeted by a voice prompt indicating the device's status, and the LCD will illuminate. Conversely, to turn the device off, rotate the same knob counterclockwise until you hear a click.

Adjusting the Volume

The volume knob also controls the loudness of the audio. Once the radio is powered on, rotating the knob clockwise increases the volume, and counter-clockwise decreases it. Fine-tune the volume to a comfortable level for your current environment.

Changing Frequencies

Baofeng radios typically operate in two modes: Frequency Mode and Channel Mode. In Frequency Mode, you can manually input a specific frequency. To do this, ensure your radio is in Frequency Mode, typically indicated by a 'Frequency' or 'VFO' displayed on the LCD. Use the keypad to enter the desired frequency. In Channel Mode, the radio accesses pre-programmed channels. You can switch between channels using the arrow keys or by keying in the channel number.

Transmitting and Receiving

To transmit, press the Push-to-Talk (PTT) button located on the side of the radio. Always remember to identify yourself with your call sign at the start and end of a transmission and periodically during a lengthy one. To receive, simply release the PTT button. You should hear incoming transmissions through the speaker, provided you're tuned to a frequency with activity and within range.

Selecting Power Level

Baofeng radios allow you to select the power level for your transmissions, which can help preserve battery life or extend the range of communication. Depending on your model, you can typically switch between low, medium, and high power settings through the menu.

Accessing the Menu

For more advanced settings, Baofeng radios offer a comprehensive menu system. Access it by pressing the 'Menu' button, then use the arrow keys or keypad to select different menu items. You can change a multitude of settings, from squelch level to power level, and more. A 'Confirm' or 'Exit' button allows you to set your choices or leave the menu.

Scanning

Scanning allows your radio to automatically cycle through frequencies or channels to find activity. This feature is useful for monitoring several frequencies or searching for signals in an unfamiliar area. Usually, you can initiate a scan by pressing a dedicated button or through the menu.

Saving Channels

If you frequently use specific frequencies, you can save them as channels for easy access. After tuning to a desired frequency, you can typically save it by accessing a 'Save' or 'Memory' function in the menu.

These are the fundamental operations for using a Baofeng radio. However, remember that each model may have slight variations, so always refer to your radio's manual for specifics. With a firm grasp of these basics, you are now equipped to delve into the more advanced aspects of your Baofeng radio, including programming and customization, which we will cover in the following chapters. It's time to turn theory into practice, and truly get to grips with the potential of your Baofeng radio.

Advanced Features

Moving beyond the basic operations of your Baofeng radio, it's time to explore the more sophisticated features at your disposal. These functions, like scanning for active frequencies, programming memory channels, and using the dual watch mode, will elevate your user experience and communication capabilities. By harnessing these features, you can adapt to changing circumstances in survival and emergency situations.

Scanning for Active Frequencies

Scanning for active frequencies allows your radio to automatically monitor a range of frequencies for any radio activity. To activate the scanning function on most Baofeng models, you need to press the 'Scan' button or select the scan function from the radio's menu. The radio will then cycle through the available frequencies or the pre-programmed channels at a swift pace. When it detects an active transmission, it momentarily stops scanning to let you listen in. If you wish to interrupt the scanning process at any point, you can usually do so by pressing the 'Scan' button again or the 'Push-to-Talk' button.

Programming Memory Channels

One of the strengths of Baofeng radios lies in their capacity to program memory channels. With this feature, you can save frequently used frequencies or channels for easy access. Once a frequency is stored, you can recall it without manually inputting the frequency each time. To program a memory channel, start by switching to Frequency Mode and entering the desired frequency. Once the frequency is set, access the menu and navigate to the 'Memory Channel' option. Here, you need to select an available memory slot for your new channel, then save the frequency to that slot. Exit the menu to finalize the process.

Some models also allow you to name your channels for even more straightforward navigation. This process varies between models, so consult your user manual for the specifics on your particular radio.

Dual Watch Mode

The dual watch mode, also known as dual standby, is a highly practical feature that enables you to monitor two frequencies or channels simultaneously. This function is particularly useful in emergency situations where you need to stay abreast of multiple communication channels.

When dual watch is enabled, your radio alternates between the two selected frequencies at regular intervals. If activity is detected on either frequency, the radio will pause on that frequency to let you hear the transmission.

To activate dual watch mode, access the menu and find the 'Dual Watch' or 'TDR' (Two Dual Reception) option. Select this feature and turn it on. The radio should now monitor both the primary and secondary frequencies or channels displayed on the screen.

Squelch Settings

The squelch setting on your Baofeng radio helps suppress unwanted background noise when there's no communication taking place. You can adjust this setting in the menu, with higher levels being more restrictive and allowing fewer weaker signals through.

VOX (Voice Operated Transmission) Function

Some Baofeng radios feature the VOX function, allowing hands-free operation. When activated, your radio will automatically transmit when it picks up your voice or other nearby sounds. You can usually adjust the sensitivity of the VOX function in the menu.

These advanced features offer remarkable versatility and efficiency, providing a robust tool for survival and emergency scenarios. Familiarize yourself with these features and practice using them regularly. In the next subchapter, we will dive deeper into customizing your Baofeng radio, enhancing its functionality even further to suit your specific needs and circumstances.

Manual Programming

Although modern technology has brought us programming software to make the task easier, understanding manual programming is crucial. You may not always have access to a computer in survival situations or emergencies. This subchapter provides a comprehensive, step-by-step guide on how to program your Baofeng radio manually, including setting frequencies, adjusting squelch levels, and saving channels.

Setting Frequencies

To manually program your Baofeng radio, you need to set frequencies first.
Here are the steps:

1. Ensure the radio is in **'Frequency Mode'**. If not, switch from **'Channel Mode'** to **'Frequency Mode'** by pressing the **'VFO/MR'** button.
2. Input the frequency you want to use. Use the numeric keypad to do this. For instance, to set the frequency to **146.520 MHz**, you would press **1-4-6-5-2-0**. If you make a mistake, the **'*'** button usually serves as a delete key.
3. Press the **'Confirm'** or **'# key'** to save your frequency.

Setting Squelch Levels

Squelch is used to mute the speaker when no signal is really strong enough to be clear.
Here's how to adjust the squelch level on your radio:

1. Press **'Menu'** and then **'0'**. This will bring up the squelch menu.

2. Use the **up** and **down** arrows to adjust the squelch level. A higher squelch level will require a stronger signal to break the squelch (mute). Be aware that a squelch level that's too high might block out weaker signals.

3. Press 'Menu' again to set the squelch level, and 'Exit' to leave the menu.

Saving Channels

After setting your preferred frequency and squelch levels, you need to save these settings to a memory channel:

1. Press the 'Menu' button followed by '2-7'. This will take you to the 'Memory Channel' menu.
2. Press 'Menu' again. The display will flash and prompt you to select a channel number.
3. Using the up and down arrow keys, choose the memory channel you want to use. Make sure not to overwrite a pre-existing channel.
4. Press 'Menu' to save the settings to that channel.
5. Exit the menu by pressing 'Exit'.

Remember, each Baofeng radio model might have slightly different procedures and button labels. Always refer to your user manual for exact instructions.

As a word of advice, label and organize your channels logically, making it easier for you to remember and access them in an emergency.

Manual programming might seem daunting initially, especially for beginners. However, with regular practice, the process becomes straightforward and instinctive. You'll be setting frequencies, adjusting squelch, and saving channels like a pro.

In the next subchapter, we'll delve into programming your Baofeng radio using software. While manual programming is essential for understanding and flexibility, software programming offers a faster and more efficient alternative, especially when dealing with a large number of channels or complex configurations.

Programming with CHIRP

CHIRP is an open-source software designed to make the programming process more intuitive and efficient. It is compatible with a wide variety of radios, including various Baofeng models. In this subchapter, we'll walk through the process of programming your Baofeng radio using CHIRP, from downloading the software to importing and uploading frequency lists.

Downloading and Installing CHIRP
The first step is to download the CHIRP software onto your computer:
1. Visit the CHIRP website (http://chirp.danplanet.com) and navigate to the 'Get It' section.
2. Select the version compatible with your operating system (Windows, MacOS, or Linux).
3. Follow the on-screen instructions to download and install the software.

Please note that you might need to adjust your computer's security settings to allow the software to download and install, as it's open-source and may not be recognized by your operating system's security protocols.

Connecting Your Radio to Your Computer
To program your radio using CHIRP, you will need to connect your radio to your computer. This usually requires a specific type of cable (often a USB programming cable):
1. Turn off your radio and connect it to your computer using the programming cable.
2. Make sure the cable is fully inserted into the radio's accessory port. Improper connections can cause errors during the programming process.
3. Turn on the radio.

Importing and Uploading Frequency Lists
With CHIRP installed and your radio connected, you're now ready to import and upload frequency lists:
1. Open the CHIRP software and click on the 'Radio' tab in the menu. Select 'Download from Radio'.

2. A box will appear asking you to choose your port, vendor, and model. For the port, select the one that corresponds to your programming cable. The vendor is Baofeng, and the model will be the specific model of your radio.
3. Click 'OK'. CHIRP will download the current contents of your radio. This may take a few minutes.
4. Once the download is complete, you'll see a spreadsheet-like interface displaying all the channels currently programmed into your radio. You can modify existing entries or add new ones.
5. To import a frequency list, go to 'File' and click 'Import'. You can then select your frequency list file. It will append to the existing list.
6. When you're done making changes, click on 'Radio' and then 'Upload to Radio'. CHIRP will send your new configuration to the radio.

Remember, every time you modify the frequency list in CHIRP, make sure to save your work. Saving the file gives you a backup that you can revert to if needed.

Programming your Baofeng radio using CHIRP can simplify and speed up the process, especially if you have a large number of frequencies to program. However, it's also essential to be familiar with manual programming for situations where a computer may not be accessible.

In the next subchapter, we'll discuss practical tips and tricks for optimizing your Baofeng radio experience, ensuring you get the most out of your device in any survival or emergency situation.

Customizing Settings

One of the advantages of Baofeng radios is their versatility and customizability. There are various settings you can adjust to optimize the radio for your needs and preferences. This subchapter will explore some of these options, including the roger beep, timeout timer, and battery saver.

Roger Beep

The roger beep is a tone that is emitted at the end of transmission to signify that the user has finished speaking. It's traditionally used in radio communication as a clear signal that the channel is now free for others to use.

Here's how you can enable or disable the roger beep:
1. Turn on your Baofeng radio.
2. Press the 'Menu' button.
3. Use the up and down arrow keys to navigate to 'ROGER'.
4. Press 'Menu' again to select this option.
5. Use the up and down arrows to turn the roger beep ON or OFF.
6. Press 'Menu' to confirm your choice.
7. Press 'Exit' to return to the main screen.

Timeout Timer (TOT)

The Timeout Timer is a safety feature that prevents continuous transmission over a set period, thus avoiding potential damage to the radio.

By default, this is typically set at 600 seconds (10 minutes), but you can adjust it as follows:
1. Press the 'Menu' button.
2. Use the arrow keys to find the 'TOT' option.
3. Press 'Menu' to select it.
4. Use the arrow keys to set the timeout duration (in seconds).
5. Press 'Menu' to confirm your choice.
6. Press 'Exit' to return to the main screen.

Please note that continuous transmission for extended periods can overheat the radio and damage the internal components.

Battery Saver

The battery saver function is a power-saving feature that, when enabled, puts the radio into a low-power mode during periods of inactivity. It's particularly useful in survival and emergency situations when conserving battery life is essential. To adjust the battery saver settings:

1. Press the 'Menu' button.
2. Use the arrow keys to find the 'SAVE' option.
3. Press 'Menu' to select it.
4. Use the arrow keys to choose between OFF and levels 1-4 (with 4 being the highest power-saving level).
5. Press 'Menu' to confirm your choice.
6. Press 'Exit' to return to the main screen.

Voice Operated Transmit (VOX)

VOX is a feature that enables your Baofeng radio to transmit voice without pressing the Push to Talk (PTT) button. The radio will automatically begin transmitting when it picks up your voice or any sound near the microphone. This function can be handy when you need hands-free operation, especially in a survival situation. To set up VOX:

1. Press 'Menu'.
2. Use the arrow keys to find 'VOX'.
3. Press 'Menu' again to select.
4. Use the arrow keys to select the desired VOX level (OFF, or 1-10, with 10 being the most sensitive).
5. Press 'Menu' to confirm.
6. Press 'Exit' to return to the main screen.

Remember that a higher VOX level will make the radio more sensitive to sound, so ambient noise could trigger transmission.

Dual Watch (TDR)

Dual Watch is another feature of Baofeng radios that allows the radio to monitor two channels simultaneously. You can respond to transmissions on either channel at any given time. To enable Dual Watch:

1. Press 'Menu'.
2. Use the arrow keys to find 'TDR'.
3. Press 'Menu' again to select.
4. Use the arrow keys to turn Dual Watch ON or OFF.
5. Press 'Menu' to confirm.
6. Press 'Exit' to return to the main screen.

Frequency Step (STEP)

Frequency Step determines the increment/decrement in frequency when you're manually adjusting it using the arrow keys. You can choose from a range of values for more granular control over frequency adjustment. To set the Frequency Step:

1. Press 'Menu'.
2. Use the arrow keys to find 'STEP'.
3. Press 'Menu' again to select.
4. Use the arrow keys to choose the desired step size (e.g., 2.5kHz, 5kHz, 6.25kHz, 10kHz, 12.5kHz, 20kHz, 25kHz, and 50kHz).
5. Press 'Menu' to confirm.
6. Press 'Exit' to return to the main screen.

Wide/Narrow Bandwidth (WN)

This setting allows you to switch between wide and narrow bandwidths. A wide bandwidth (wideband) provides better sound quality but takes up more spectrum space, while a narrow bandwidth (narrowband) conserves spectrum space but might have lesser sound quality. To switch between wide and narrow bandwidths:

1. Press 'Menu'.
2. Use the arrow keys to find 'WN'.
3. Press 'Menu' again to select.
4. Use the arrow keys to choose between Wide (W) and Narrow (N).
5. Press 'Menu' to confirm.
6. Press 'Exit' to return to the main screen.

These are just some of the settings you can customize on your Baofeng radio. Other options include adjusting the squelch level, setting the display illumination duration, and enabling or disabling the dual watch mode. Understanding these features and knowing how to adjust them can significantly enhance your user experience, especially in emergency or survival situations.

In the following subchapter, we'll explore some practical tips for maintaining your Baofeng radio, which will help ensure its longevity and reliability when you need it most.

Chapter 3: Staying Safe With Your Baofeng

Emergency Communication Protocols

When it comes to emergency situations, the adage "be prepared" rings especially true. From natural disasters to sudden crises, efficient communication can mean the difference between life and death. Given that cellular networks may not always be reliable or available in emergencies, handheld radios like the Baofeng models become indispensable for their robustness and independence from such networks.

This chapter connects the knowledge you've gained about customizing your Baofeng radio settings in the last subchapter with the crucial emergency communication protocols that should be observed in a crisis scenario.

Calling Frequencies

Starting off with the calling frequencies - these are pre-determined frequencies meant to initiate contact with other radio operators. When using your Baofeng radio, remember to set your device to the appropriate calling frequency for your region. For instance, in the United States, the nationally recognized calling frequency on the 2-meter band is 146.520 MHz. This frequency is generally monitored by radio operators and is also the frequency you should use to signal for help in an emergency.

Operating your radio on a calling frequency involves the following steps:
1. Transmit your call sign and mention that you're listening or "monitoring". For example, "This is [your callsign], monitoring."
2. Pause and listen for any responses from other stations.
3. If you get a response and wish to continue communication, suggest switching to a different frequency to keep the calling frequency clear for others.

Note: Avoid long conversations on calling frequencies. The purpose of these frequencies is to establish initial contact before moving to a less crowded frequency for more extended communication.

Distress Signals

Once you've grasped the calling frequencies, it's crucial to understand distress signals. These signals are used to convey urgent emergencies where there is immediate danger to life or property. The universally accepted distress signals include **'MAYDAY'** for life-threatening emergencies and **'PAN-PAN'** for situations that are urgent but not immediately life-threatening. Activating a distress signal involves the following steps:

1. Set your radio to the calling frequency that's most likely to be monitored in your area.
2. Broadcast the appropriate distress signal followed by your call sign, your location, and the nature of the emergency.

Remember: Use the 'MAYDAY' signal only in life-threatening situations. Misuse can result in legal penalties.

Phonetic Alphabet

In high-stress situations or when signal quality is poor, it's important to ensure that messages are received and understood correctly. That's where the phonetic alphabet comes in. This is a set of codewords used to represent each letter of the English alphabet. It helps to eliminate misunderstandings that can arise from letters that sound similar when spoken. For instance, if you are communicating your call sign and it includes the letter 'D', using the phonetic alphabet, you would say 'Delta' instead. The phonetic alphabet word for 'D' is 'Delta'.

This is an essential communication protocol that you need to learn for effective radio communication. Familiarize yourself with the complete phonetic alphabet, starting from **'Alpha' for 'A'**, **'Bravo' for 'B'**, all the way to **'Zulu' for 'Z'**. Practice transmitting and understanding this alphabet to ensure seamless communication during emergencies.

By integrating your knowledge of Baofeng radio settings with these communication protocols, you can be better prepared for any emergency situation that might arise. In the following subchapter, we'll delve deeper into the practical aspects of radio communication in emergencies.

Connecting with Emergency Services

In the previous subchapter, we explored the basics of emergency communication protocols such as calling frequencies, distress signals, and the phonetic alphabet. Now, we delve deeper into how you can connect directly with emergency services using your Baofeng radio.

It's important to note that connecting with emergency services using a Baofeng radio should be a last resort when no other means of communication is available. It's critical to use this radio responsibly to prevent unnecessary disruption to emergency channels that could delay help to those in real need.

Identifying the Right Frequencies

The first step to connect with emergency services is knowing the correct frequencies. Emergency services often operate on specific frequencies reserved for their use. These can differ depending on the country and even region you're in. In the United States, many emergency services, including police, fire, and EMS, operate on a variety of frequencies in the VHF and UHF bands.

However, it's crucial to remember that the Baofeng radios are typically not designed for direct communication with these frequencies. Transmitting on these frequencies without appropriate authorization is generally illegal and could lead to penalties.

On the other hand, there are frequencies that can be monitored and used by the public in emergencies, like the Citizens Band (CB), Family Radio Service (FRS), and General Mobile Radio Service (GMRS). The National Weather Service (NWS) also broadcasts on specific frequencies, providing weather updates and emergency information, which can be received on some Baofeng models.

Communicating Effectively

Assuming you're in a dire situation and need to reach out to emergency services, communicating effectively is essential. Here are some steps to follow:

1. **Calm and Clear Communication:** Despite the emergency situation, remain calm while communicating. Clearly state your situation, location, the help needed, and any other relevant information. Avoid unnecessary chatter as it could interfere with other emergency communications.

2. **Use the Phonetic Alphabet:** As mentioned in the previous chapter, use the phonetic alphabet for spelling out important information. This ensures the message is understood correctly, even in challenging reception conditions.

3. **Listen More, Talk Less:** Always remember that radio communication is a two-way process. After communicating your emergency, listen for any instructions or questions.

4. **Use Standard Protocols:** When initiating the communication, use standard protocols. Start with the distress signal 'MAYDAY' (for life-threatening emergencies) or 'PAN-PAN' (for urgent but not immediately life-threatening situations) followed by your call sign, your location, and the nature of the emergency.

5. **Regular Updates:** If help is delayed, give regular updates about your situation. If your situation improves or worsens, ensure you communicate that too.

Respecting Regulations

While a Baofeng radio can be a lifeline in an emergency, it's important to respect regulations surrounding its use. Remember, unauthorized transmission on emergency service frequencies can lead to severe penalties. Therefore, always use the radio responsibly and within the bounds of the law.

Know the Laws

Before you even turn on your Baofeng radio, it's vital to understand the laws regulating its use in your area. As mentioned earlier, transmitting on frequencies allocated to emergency services without authorization is illegal in most places. Each country has its own rules regarding this, so familiarize yourself with the laws in your region.

Locate the Frequencies

When you need to use your radio in an emergency situation, knowing the frequencies that you can legally use is crucial. Often, these frequencies are reserved for amateur radio operators or general use. In the U.S., for example, the FRS and GMRS frequencies can be used without the need for a license, although GMRS frequencies may have power restrictions. It's a good idea to save these frequencies in your radio's memory channels so you can quickly access them in an emergency.

Prepare for Communication

Before reaching out, take a deep breath and collect your thoughts. You need to be able to communicate your situation quickly and effectively. Make sure to state your location, the nature of your emergency, and the type of assistance you need.

Step-by-Step Guide to Transmitting a Distress Call

Here is a step-by-step guide for transmitting a distress call:

1. **Select the Appropriate Frequency:** Start by selecting the appropriate frequency for emergency communication. It could be a designated distress frequency or a frequency that you know is regularly monitored, such as the NOAA Weather Radio frequencies in the United States.
2. **Start with a Distress Signal:** Begin your transmission by repeating the appropriate distress signal three times. If you are in immediate danger, use 'MAYDAY.' If you require assistance but are not in immediate danger, use 'PAN-PAN.'
3. **State Your Call Sign and Location:** After transmitting the distress signal, state your call sign, if you have one, followed by your precise location. If possible, use GPS coordinates.
4. **Describe the Nature of the Emergency:** Clearly describe the emergency situation and specify the kind of assistance you need.
5. **Wait for a Response:** Once you have transmitted your distress call, wait for a response. Remember, it's important to listen more than you speak.
6. **Repeat If Necessary:** If you do not receive a response, repeat your distress call. Remember to keep your message concise and clear.

This process underscores the importance of learning how to operate your Baofeng radio correctly and understanding the principles of radio communication. The more comfortable you are with your radio, the more effectively you'll be able to use it when it matters most. The next subchapter will delve into participating in amateur radio emergency services, further building on your skills and knowledge.

Building a Survivalist Network

Having understood the rudiments of emergency communication protocols and established how to connect with emergency services, we will now delve into the concept of building a survivalist network. This involves forming and maintaining a network of other survivalist radio operators. It is an essential part of your preparedness for survival situations, as it can greatly extend your communication reach and resources.

Why Build a Survivalist Network?

Isolation can be a huge detriment in an emergency situation. Being part of a network of like-minded survivalists, however, offers a shared wealth of knowledge, resources, and support. This is where ham radio comes into play. With it, you can build a network that stretches across towns, regions, or even countries. This network can be leveraged for information exchange, assistance, or moral support during challenging times.

Finding Like-Minded Operators

The first step in building your network is to find other survivalist radio operators. A good place to start is local amateur radio clubs. These clubs usually consist of enthusiasts who appreciate the technical aspects of radio communication, and among them, you'll often find individuals who share your interest in survivalism.

Online forums and social media groups are another valuable resource. Websites like QRZ.com and groups on platforms like Facebook or Reddit often have sections dedicated to survivalist radio operation. These communities can provide advice, guidance, and potential contacts.

Establishing Communication Protocols

Once you've started to build your network, it's important to establish clear communication protocols. This can include designated check-in times, codes for specific situations, or agreed-upon frequencies for different types of communication.

For instance, you might have a weekly check-in time where all network members turn on their radios and confirm their status. You might also have a specific code to indicate a distress situation, separate from the standard 'MAYDAY' or 'PAN-PAN.' This would inform network members of an emergency without necessarily alerting any unintended listeners.

Coordinating Frequencies

In addition to communication protocols, coordinating frequencies among your network is essential. Each member should have a list of frequencies used by the network, along with their specific purposes. This might include a frequency for general communication, another for emergencies, and yet another for data transmission.

One practical way to organize this is to create a shared document, like a Google Spreadsheet, which all members can access and update. This could include columns for the frequency, its purpose, the time it's typically active, and any other relevant notes.

Maintaining the Network

Maintaining your survivalist network requires regular communication, training, and updates to your agreed-upon protocols and frequencies. Check-ins not only confirm that everyone is safe but also ensure that everyone's equipment is working properly. Regular training sessions, whether in-person or over the radio, can also be a great way to ensure that all members are comfortable with the protocols and ready to respond in an emergency. Building a survivalist network is not an overnight task. It requires time, commitment, and consistency. However, the benefits of having a group of individuals you can rely on in an emergency are worth the effort. In the next section, we will look at troubleshooting common issues, rounding off your skill set as a well-prepared survivalist radio operator.

Caring for Your Radio

While forming a survivalist network significantly improves your preparation for emergencies, the central pillar of your communication resilience remains your ham radio itself. Therefore, caring for your radio is of paramount importance. Keeping it in good working order ensures that it will function reliably when you need it most. This subchapter will explore best practices for maintaining your radio, including cleaning, battery care, and proper storage.

Cleaning Your Radio
A well-kept radio starts with cleanliness. Dust and grime can get into the device's crevices, impacting its performance over time. To clean your radio, use a soft, dry cloth to wipe the outer casing. Avoid using liquids, as they can seep into the radio and damage the internal components. For the tiny corners and buttons, use a dry soft-bristle brush or canned air to gently remove any accumulated dust.

Taking Care of the Battery
Your radio's battery is its life source, and its care is crucial. Firstly, always use the battery recommended by the manufacturer to avoid damage to the radio. It's also essential to charge the battery fully before using the radio for the first time. This conditions the battery for future recharges and helps it reach its full capacity.

For ongoing battery maintenance, it's recommended not to wait until the battery is completely drained before charging. Charge the battery when it's around 20% to prolong its life. It's also advisable to remove the battery if you won't be using the radio for an extended period. When storing batteries, keep them in a cool, dry place as extreme temperatures can degrade their performance and lifespan.

Safe Storage of Your Radio
When you're not using your radio, how and where you store it can significantly impact its longevity. Ideally, store your radio in a dry, temperature-controlled environment. Extreme heat can damage the radio's internal components, while excessive cold can impair battery performance. A storage case can protect your radio from dust and physical damage. If you're storing your radio for an extended period, it's a good practice to remove the battery and antenna to prevent potential damage.
If your radio comes into contact with moisture, avoid turning it on immediately as this may lead to a short circuit. Instead, remove the battery and antenna and let the radio dry out completely in a warm, dry place before attempting to use it again.

Periodic Checks

Apart from the routine maintenance practices, periodic checks and tests are an integral part of radio care. Regularly power up your radio and check all functions to ensure they're working as they should. This includes dialing through different frequencies, checking the clarity of transmitted and received signals, and verifying the condition of the battery.

Dealing with Wear and Tear

Despite meticulous care, your radio may show signs of wear and tear over time. Buttons might become less responsive, or the display might not be as bright as it used to be. In such cases, it's essential to consult the radio's manual or contact the manufacturer's customer service. Often, these issues can be resolved with simple fixes, ensuring your radio continues to serve you for years to come.

In conclusion, proper care and maintenance of your radio are just as important as knowing how to use it. Next, we'll focus on understanding and troubleshooting common issues you may encounter, arming you with the knowledge to keep your radio at its peak performance.

Common Problems and Solutions

Building on the fundamentals of radio care from the previous subchapter, we now delve into addressing common problems encountered by Baofeng radio users. Radios are sophisticated devices, and sometimes, they may not function as expected. Inability to transmit, poor reception, and programming issues are among the common challenges. However, with a little know-how, most of these problems can be diagnosed and resolved right at home.

Problem 1: Inability to Transmit
If you find yourself unable to transmit signals from your Baofeng radio, there are a few things to check. The first step is to verify that you have correctly set the frequency, which can be checked in your radio's manual or on the LCD screen of your device. Ensure that you're on the correct mode for the band you're attempting to use. If you're within the proper band limits and the problem persists, inspect your antenna. Ensure it is securely attached and in good condition. Damaged antennas can severely impair your radio's ability to transmit.

Problem 2: Poor Reception
Another common issue with radios is poor reception, characterized by weak or distorted signals. This issue can be due to several factors. Start by checking the signal strength on your radio's display. If it's low, try changing your location to somewhere with fewer physical obstructions. Buildings, mountains, and even dense foliage can interfere with signal reception.
In some cases, poor reception could be due to an issue with the antenna. Check to ensure that the antenna is properly attached and isn't damaged. If the antenna looks worn out or damaged, it may need replacement. Lastly, the radio's squelch setting might be set too high. Lowering the squelch level might bring in the desired signal; however, note that this could also let in more background noise.

Problem 3: Issues with Programming
Programming can seem daunting, especially for beginners, and it's not uncommon to encounter issues in this area. If you're struggling with manual programming, ensure you're following the correct steps as outlined in the radio's manual or in the previous subchapter of this book. Double-check each step, as a single misstep can lead to programming errors.
If you're using CHIRP software for programming, make sure you've installed the correct drivers for your programming cable. Without the right drivers, your computer may not recognize the device, leading to programming issues. Also, ensure you're using the latest version of the CHIRP software, as older versions may have bugs or compatibility issues.

Problem 4: Hearing Unwanted Conversations or Interference

One of the potential challenges that you might face is hearing unwanted conversations or interference on your desired frequency. This is generally because of a weak squelch setting or interference from nearby frequencies. If the squelch setting is too low, your radio may pick up weak signals that can include static, noise, or other conversations. To address this, you can increase the squelch level in your radio settings.

Conversely, if the interference is due to nearby frequencies, it's essential to understand that this can be a common issue in densely populated areas with many active radio users. One possible solution is to adjust your frequency slightly. However, you must ensure that the frequency you select is legal and appropriate for your license class and communication purpose.

Problem 5: Difficulty Connecting to Repeaters

Connecting to repeaters can sometimes prove challenging for new operators. If you're having difficulty connecting to a repeater, there could be several causes. Firstly, verify that you have the correct frequency, offset, and tone for the repeater. These details are usually available on the repeater operator's website or through local radio clubs.

Next, ensure that your radio is set up to use the correct offset direction. A wrong offset setting can prevent you from accessing the repeater. Furthermore, make sure that you are within range of the repeater. Repeaters increase the range of your signal but they do have limits, and these limits can be affected by geographical factors like hills or buildings.

Problem 6: Radio Suddenly Stops Working

If your radio stops working suddenly, begin by checking the most obvious: the battery. Ensure that your battery is adequately charged. Remember, transmitting requires more power than receiving, so even a battery with a partial charge may be insufficient for transmitting.

If your battery is fully charged but the radio still doesn't work, try doing a factory reset. Keep in mind that performing a factory reset will erase any programmed channels or custom settings, so it's a good idea to keep a backup of your radio's configuration.

In all of these scenarios, if the problems persist despite your best efforts at troubleshooting, it may be necessary to consult with a more experienced operator or a professional. Remember, it's a learning process, and encountering and overcoming these common issues is all part of becoming proficient with your Baofeng radio. As we proceed in the following subchapter, we will explore how accessories can be used to enhance your radio's functionality and circumvent some common issues.

Chapter 4: Case Studies and Practical Exercises

Real-Life Use Cases

Baofeng radios, with their vast capabilities and affordability, have proven themselves invaluable tools in numerous real-life scenarios. In this subchapter, we will delve into specific instances where these radios have been used effectively in survival or emergency situations, underscoring their practicality and potential for life-saving communication.

Case 1: Natural Disaster - Hurricane Katrina

When Hurricane Katrina struck the Gulf Coast of the United States in 2005, it caused devastating damage and resulted in a significant loss of life. In the aftermath, many found themselves isolated from their families and emergency services, as the storm had severely disrupted traditional communication systems like landlines and cell towers. In such a situation, a number of individuals with Baofeng radios were able to maintain contact with each other and with emergency services. The high-frequency bands and repeater access of these radios enabled communication over large distances, often being the only reliable means of connection in a shattered communications landscape.

Case 2: Outdoor Recreation - Hiking and Mountaineering

Baofeng radios have proven their worth in the realm of outdoor adventure as well. Consider the case of a group of mountaineers climbing a high peak in the Rocky Mountains. When a sudden snowstorm separated the team, they were able to use their Baofeng radios to maintain contact, relay their positions, and eventually regroup safely. Their cell phones were ineffective in the remote area, but the Baofeng radios, with their robust range and terrain-penetrating VHF and UHF frequencies, kept the team connected and played a crucial role in their safe return.

Case 3: Community Event - City Marathon

At a large-scale event like a city marathon, coordination and communication among event staff and volunteers are essential. One such event utilized Baofeng radios to maintain clear, instant communication throughout the race course. The radio's ability to communicate over a wide area and on multiple channels allowed for efficient coordination, rapid response to any issues or emergencies, and ultimately contributed to the smooth running of the event.

Case 4: Rural Living - Farming Communities

In rural areas where cell phone signals can be unreliable or non-existent, Baofeng radios have also demonstrated their utility. In one instance, a farming community used these radios as their primary means of communication. The radios allowed farmers in expansive fields to stay connected with their homes and with each other, offering not only convenience but also safety in case of an emergency.

Case 5: Fire Emergency in Residential Area

In a dense suburban neighborhood, a fast-spreading fire forced residents to quickly evacuate. Amidst the chaos, a group of neighbors with Baofeng radios was able to maintain communication with each other and with local emergency services, coordinating safe exit routes and accounting for every resident. In a situation where cell phone networks were overwhelmed, the radios were an essential lifeline.

Case 6: Power Outage in Urban Area

A major city faced an extensive power outage that lasted several days. Not only did this situation disrupt daily life, but it also rendered many conventional communication devices useless. Citizens with Baofeng radios managed to stay informed and connected, coordinating shared resources and relaying updates about the situation.

Case 7: Search and Rescue Operation

A young hiker got lost in a remote national park. Search and rescue teams, aided by volunteers using Baofeng radios, were able to cover vast areas and communicate in real-time to effectively coordinate the search operation. The hiker was eventually found and rescued, thanks in large part to the expansive communication network facilitated by these radios.

Case 8: Car Breakdown in Remote Location

A family traveling through a remote part of the country had their car break down late at night, with no cellular signal in the vicinity. Using their Baofeng radio, they were able to reach out to a local amateur radio operator, who then contacted a tow service on their behalf, demonstrating the radio's utility in everyday emergencies.

Case 9: Disaster Response Team

In a cyclone-hit coastal region, a disaster response team relied heavily on Baofeng radios to coordinate relief efforts. Their ability to operate on multiple channels enabled different groups - medical teams, food distribution volunteers, and security personnel - to function seamlessly and respond effectively to the crisis.

Case 10: Amateur Radio Community Event

At a local "Hamfest," a convention of amateur radio enthusiasts, Baofeng radios were used for wide-area communication, announcements, and event coordination. Despite hundreds of attendees over a large fairground, the event went off without a hitch, thanks to the efficient communication facilitated by these radios.

Case 11: High School Field Trip

On a school field trip to a vast national park, teachers and chaperones used Baofeng radios to keep track of students and coordinate activities. The radios ensured a constant line of communication across the park's expansive, cell-signal-starved area, keeping the trip safe and organized.

Case 12: Neighborhood Watch

A neighborhood watch group utilized Baofeng radios to maintain constant communication during nightly patrols. The radios' reliability and range proved much more effective than cell phones, enhancing the group's efficiency and the overall security of the neighborhood.

Case 13: Boating Expedition

During a boating expedition in coastal waters, a group of sailors used their Baofeng radios to maintain contact with coastal stations and other boats. When one of the boats experienced engine trouble, a distress signal was promptly sent out and assistance arrived without delay.

Case 14: Road Construction Crew

A road construction crew working in a remote location used Baofeng radios for onsite communication, coordinating machinery movement, shift changes, and safety protocols. This helped prevent accidents, streamline work, and maintain overall site safety.

Each of these instances underscores the versatility and dependability of Baofeng radios across a spectrum of scenarios, highlighting their relevance and importance in today's interconnected world.

Practical Exercises

In the journey of mastering your Baofeng radio, it's important to not just understand its operation theoretically, but also practically. This section comprises a set of carefully designed exercises that will guide you step-by-step through various features and functions, allowing you to gain hands-on experience.

Exercise 1: Simple Transmission and Reception
Objective: To establish a basic understanding of how to transmit and receive signals.
1. Ensure that your radio is properly charged.
2. Select a frequency within the allowable range, making sure it is a clear frequency not in use by other services.
3. Use the Push-To-Talk (PTT) button on the side of your radio to begin transmitting. While pressing the PTT, you should speak clearly into the microphone.
4. To receive, release the PTT button and listen.

This is a simple yet fundamental task, reinforcing the basic operation of your Baofeng radio.

Exercise 2: Programming Frequencies Manually

Objective: To learn how to manually set frequencies, set squelch levels, and save channels. Follow the steps outlined in Sub-Chapter 4.1. The aim is to feel comfortable programming your radio manually without external assistance.

Exercise 3: Programming with CHIRP
Objective: To familiarize yourself with using CHIRP to program your radio.
Follow the detailed guide in Sub-Chapter 4.2, practicing how to import and upload frequency lists. This exercise will help you get acquainted with a more efficient way of programming multiple channels.

Exercise 4: Adjusting Advanced Settings
Objective: To understand how to customize your radio settings according to your preferences.

Utilize the knowledge gained from Sub-Chapter 4.3, and practice customizing various settings like roger beep, timeout timer, and battery saver.

Exercise 5: Emergency Communication Drill
Objective: To simulate emergency communication protocols.
1. Set your radio to the designated emergency frequency (as outlined in Sub-Chapter 5.1).
2. Practice sending a standard distress signal.
3. Use the phonetic alphabet to communicate vital information like your location.

Exercise 6: Building a Communication Network
Objective: To practice establishing a network with other radio operators.
1. Coordinate with a group of fellow Baofeng users (friends, family, local radio clubs).
2. Set a common frequency and establish a communication schedule.
3. Practice relaying information, status checks, and troubleshooting together.

Exercise 7: Maintenance and Cleaning
Objective: To understand and apply the best practices for caring for your radio.
Follow the instructions detailed in Sub-Chapter 6.1 to clean your radio and learn about battery care and proper storage.

Exercise 8: Troubleshooting
Objective: To familiarize yourself with resolving common issues with your radio.

Try to simulate some of the common problems (outlined in Sub-Chapter 6.2) and then follow the recommended solutions.
Remember, these exercises aim to give you a firsthand understanding of your radio's capabilities and potential. Don't be discouraged by mistakes. Instead, consider them as learning opportunities on the path to mastering your Baofeng radio. Keep practicing and, before you know it, you will navigate the radio's features with confidence and ease

Now let's look at hypothetical scenarios that could occur in emergency situations and how to initiate communication professionally, striving to be as precise and useful as possible.

Scenario 1: A lost hiker

You've become lost while hiking in a vast national park. Luckily, you have your Baofeng radio with you. Here's how you might use it:

1.1. Turn on your radio and check the battery level. If it's low, save power by turning off unnecessary functions.

1.2. Tune to a common emergency frequency like 146.520 MHz, the national calling frequency for ham radio.

1.3. Politely break into the chatter with your call sign and state that you have an emergency.

1.4. If you get a response, explain your situation concisely: "This is [call sign]. I'm lost in [name of park]. I last remember being near [last known location]."

1.5. Follow the instructions given by the responder. If advised, switch to a quieter frequency to keep the emergency one clear.

Scenario 2: During a flood

You're in your home and a sudden flood has cut off all regular means of communication.

2.1. Ensure your radio is fully charged and switch it on.

2.2. Tune to a local emergency frequency; if you don't know it, use the national calling frequency.

2.3. Broadcast a general call for help, stating your call sign and expressing urgency: "This is [call sign]. I am in immediate danger due to flooding at [your location]."

2.4. If responders require more specific information about your situation (like water levels or number of people), provide it concisely and clearly.

Scenario 3: Witnessing a car accident

You're in a remote area and witness a severe car accident. Cell service is unavailable.

3.1. Turn on your radio and tune into a local emergency frequency.

3.2. Break into the chatter, stating your call sign and the nature of the emergency.

3.3. Give clear and precise details about the location and condition of the accident: "This is [call sign]. There's been a car accident on [road name]. There are [number] of people injured."

3.4. Follow the instructions given by any responders. They may ask you to relay further information or updates about the situation.

In all these scenarios, remember to stay calm and clear in your communication. Also, don't forget to turn off your transmission after speaking to conserve battery life and free up the frequency for responses. If the situation changes or worsens, don't hesitate to update the responders with the new information.

Scenario 4: Fire in a residential building

You've spotted a fire in a building while you're out in the city and you have your Baofeng radio with you.

4.1. Ensure your safety first and move away from the immediate vicinity of the fire.

4.2. Switch on your radio, make sure it's charged, and tune into a local emergency frequency.

4.3. Send a distress signal, stating your call sign and the nature of the emergency: "This is [call sign]. There's a fire in a residential building at [location]."

4.4. Provide any additional details about the situation such as the size of the fire, if people are in danger, or if you've spotted anyone trapped.

Scenario 5: Stranded in a snowstorm

Your car has broken down during a snowstorm in a remote area. Your mobile phone has no signal.

5.1. Stay in your vehicle for warmth and safety. Turn on your Baofeng radio.

5.2. Tune to a local or national emergency frequency.

5.3. Send a distress signal stating your call sign and situation: "This is [call sign]. I'm stranded in a snowstorm at [location]."

5.4. Follow instructions from any responders and provide more information if requested, such as your vehicle description.

Scenario 6: During a power outage

A power outage has occurred in your area, and you're unable to use your landline or mobile phone.

6.1. Turn on your radio and tune into a local emergency frequency.

6.2. Broadcast your situation: "This is [call sign]. There is a power outage at [your location]."

6.3. If you require assistance, state so clearly and concisely.

Scenario 7: Stuck in a lift

You're stuck in a lift in a large building, your mobile phone has no signal.

7.1. Remain calm, switch on your radio, and tune into a local emergency frequency.

7.2. Broadcast your situation: "This is [call sign]. I'm stuck in a lift at [building location]."

7.3. Follow any instructions given by the responders, and give them any additional information if requested, such as the floor you're stuck on.

Scenario 8: Stranded in a boat

Your boat has broken down at sea, and your mobile phone has no signal.

8.1. Make sure your radio is waterproofed or in a dry place. Turn it on and tune into a maritime emergency frequency.

8.2. Send a distress signal stating your call sign and situation: "This is [call sign]. I'm stranded at sea in a boat at [rough location]."

8.3. Provide any additional details that may assist the responders in locating you, such as your boat's name and color.

Scenario 9: Injured while camping

You're camping in a remote area, and you or someone in your party is injured and needs medical attention.

9.1. Switch on your radio and tune to an emergency frequency.

9.2. Call for help stating your call sign and the nature of the medical emergency: "This is [call sign]. I'm at [location] and need medical assistance."

9.3. Provide details about the injury, the condition of the person, and any first aid already administered.

Scenario 10: During a riot or civil unrest

During an instance of civil unrest or a riot, regular communication networks may be overloaded or shut down.

10.1. Ensure your immediate safety first, and then switch on your radio and tune into a local emergency frequency.

10.2. Broadcast your situation: "This is [call sign]. There's civil unrest at [location]. I am safe."

10.3. If you need assistance to evacuate or are injured, make it clear in your communication. Remember, during all these situations, the key to effective communication is staying calm, clear, and concise. Give responders time to react and always follow any instructions given.

Most

Important radio frequencies during an emergency:

- **121.5 MHz** - International emergency frequency for distress signals and search and rescue operations.
- **155.160 MHz** - National calling frequency for emergency services and organizations in some countries.
- **27.555 MHz** - Citizen Band (CB) emergency calling frequency.
- **146.520 MHz** - Simplex emergency frequency commonly used by amateur radio operators.
- **162.550 MHz** - NOAA Weather Radio frequency for severe weather alerts and emergency information.
- **243.000 MHz** - Military emergency frequency for air traffic control and distress signals.
- **446.00625 MHz** - PMR446 emergency calling frequency used in Europe.
- **155.475 MHz** - VHF marine emergency calling frequency.
- **2182 kHz** - Medium Frequency (MF) maritime distress and calling frequency.
- **3760 kHz** - High Frequency (HF) maritime distress and calling frequency.
- **406.025 MHz** - Emergency position-indicating radio beacon (EPIRB) frequency for satellite-based distress signals.
- **156.800 MHz** - VHF marine channel 16, commonly monitored by emergency responders.

- **122.800 MHz** - Unicom frequency for non-towered airports during emergency situations.
- **462.675 MHz** - FRS/GMRS emergency calling frequency.
- **7.110 MHz** - Amateur Radio Emergency Service (ARES) frequency for disaster communications.
- **14.300 MHz** - International Amateur Radio Union (IARU) emergency frequency on HF bands.
- **446.16875 MHz** - Emergency calling frequency for DMR (Digital Mobile Radio) systems.
- **27.185 MHz** - Single Sideband (SSB) CB emergency calling frequency.
- **146.460 MHz** - MURS (Multi-Use Radio Service) emergency calling frequency.
- **462.650 MHz** - GMRS emergency calling frequency.